Table Of Contents

From the author 2
Chapter 1: Introduction to Performance Tuning 2
    The Importance of Performance Tuning 2
    Overview of Subaru Models and Their Potential 2
    Understanding the Subaru Engine Architecture 2
Chapter 2: Setting Goals for Your Build 2
    Defining Performance Objectives 2
    Street vs. Track Performance 2
    Budgeting for Modifications 2
Chapter 3: Basic Modifications 2
    Intake Systems 2
    Exhaust Upgrades 2
    ECU Reflashing and Tuning 2
Chapter 4: Advanced Engine Modifications 2
    Turbocharger Upgrades 2
    Intercooler Systems 2
    Fuel System Enhancements 2
Chapter 5: Suspension and Handling 2
    Upgrading Shocks and Struts 2
    Sway Bars and Chassis Stiffening 2
    Tire and Wheel Selection 2
Chapter 6: Transmission and Drivetrain Modifications 2
    Upgrading Clutches and Flywheels 2
    Differential Enhancements 2
    Short Shifters and Gear Ratios 2
Chapter 7: Tuning for Performance 2
    Understanding Dyno Testing 2
    Fuel Mapping and Timing Adjustments 2
    Troubleshooting Common Issues 2
Chapter 8: Safety Considerations 2
    Essential Safety Equipment 2
    Engine and Component Reliability 2
    Legal Considerations in Tuning 2
Chapter 9: Community and Resources 2
    Engaging with the Subaru Community 2
    Online Resources and Forums 2

    Attending Events and Meets 2
Chapter 10: The Future of Subaru Performance Tuning 2
    Emerging Technologies and Trends 2
    Electric and Hybrid Performance Tuning 2
    Final Thoughts on Your Subaru Journey 2
    Prologue 1

Prologue

The author of this book is currently modifying his 2003 Subaru WRX. He hand built the engine with forged internals and has built several custom turbos for the engine. He has never been completely happy with the low end torque in the EJ20 series engine family. As part of his journey, he is leaving the COBB platform for a Link ECU. The goal is to support additional modifications down the road. The Link ECU will support water/meth injection, comprehensive boost control, wideband monitoring, map vs e-map pressure monitoring, and much, much more. He has purchased a Scorpion Supercharger Solutions adapter that will allow him to install an Eaton M45 roots style supercharger from a Mini Cooper. The plan for this arrangement is to have the supercharger handle off-idle boost until the turbo can spool up before gradually ramping up boost as RPM increases. With this modification along with a modified twin-scroll turbocharger and bellmouth downpipe to 3 inch exhaust, 1000cc Injector Dynamics ID1000 fuel injectors, estimations on power production will require upgrades to the differential(s) and transmission. He plans to install a six speed X-Shift sequential gearbox (helical) along with LSD differentials.

From the author

This book is meant to be a basic introduction to modification and upgrade paths. If the response is great enough, I will write detailed follow-ups. Thank you for joining me on my journey, please enjoy!

You can observe my journey at:
https://www.youtube.com/@SuperchargetheWorld
You can support my company and help launch my product here: https://www.kickstarter.com/profile/scorpionsolutions

Chapter 1: Introduction to Performance Tuning

The Importance of Performance Tuning

Performance tuning plays a pivotal role in the automotive modification landscape, particularly for enthusiasts dedicated to Subaru vehicles. As modders and tuners delve into the intricacies of their cars, understanding the importance of performance tuning becomes essential. This process involves optimizing the engine's settings to achieve maximum power, efficiency, and overall vehicle performance. By fine-tuning various components, from the air-fuel mixture to ignition timing, enthusiasts can unlock their Subaru's full potential, transforming it from a factory model into a powerful machine that can handle the demands of both the road and the track.

One of the primary benefits of performance tuning is the enhancement of engine efficiency. Subaru engines, renowned for their boxer configuration, can be further optimized through careful tuning. By adjusting parameters such as fuel delivery and boost levels, tuners can achieve a more complete combustion process. This not only leads to increased horsepower but also improves fuel economy. For automotive modders, this balance between performance and efficiency is crucial, especially for those who use their vehicles for daily driving alongside spirited weekend escapades.

Moreover, performance tuning allows for the customization of the driving experience. Each tuner has a unique vision of what their ideal vehicle should be, and tuning provides the means to bring that vision to life. Whether it's achieving a smoother throttle response, enhancing turbocharger performance, or adjusting the suspension settings, performance tuning enables Subaru enthusiasts to tailor their vehicles to their specific preferences. This personalization fosters a deeper connection between the driver and their car, making the experience more enjoyable and engaging.

Additionally, performance tuning contributes to vehicle reliability and longevity. Contrary to the common misconception that tuning always leads to higher stress on engine components, when done correctly, tuning can enhance the durability of an engine. By ensuring that the engine operates within its optimal parameters, tuners can prevent issues such as knocking or overheating. This proactive approach not only maximizes performance but also extends the lifespan of the vehicle, making it a worthwhile investment for modders who wish to enjoy their Subaru for years to come.

Finally, the importance of performance tuning extends beyond individual vehicles; it plays a significant role in the broader automotive community. As Subaru enthusiasts share their tuning experiences, techniques, and results, they contribute to a culture of knowledge and innovation. This exchange of information fosters a collaborative environment where tuners can learn from one another, pushing the boundaries of what is possible in automotive performance. By embracing the art and science of performance tuning, Subaru modders not only elevate their own experiences but also help to advance the entire tuning community. In this way, performance tuning becomes not just a means to an end, but a vital component of the shared passion that unites Subaru enthusiasts worldwide.

Overview of Subaru Models and Their Potential

In the world of automotive enthusiasts, Subaru holds a distinctive place, renowned for its unique blend of performance, reliability, and versatility. The brand's models are not just vehicles; they are platforms that invite modifications and tuning, making them popular among automotive modders and tuners. This subchapter provides an overview of the various Subaru models, highlighting their inherent potential for performance enhancements and the aftermarket modifications that can elevate their capabilities.

Starting with the iconic Subaru Impreza, particularly the WRX and WRX STI variants, these models have become synonymous with rally-inspired performance. The turbocharged flat-four engines, coupled with Subaru's renowned all-wheel-drive system, offer a robust foundation for tuning. With a strong aftermarket support network, enthusiasts can explore a range of modifications, from simple ECU remapping to more complex upgrades like larger turbochargers and performance exhaust systems. The potential of the Impreza lies not only in its raw power but also in its agility and handling, making it a favorite among those seeking spirited driving experiences.

Moving on to the Subaru BRZ, this model presents a different tuning challenge. As a lightweight sports coupe, the BRZ is known for its balanced chassis and rear-wheel-drive layout. While it may not boast the turbocharged power of its siblings, the BRZ's potential lies in its handling dynamics and responsiveness. Modders often focus on suspension upgrades, lightweight wheels, and performance-oriented exhaust systems to unlock the BRZ's full potential. Additionally, enthusiasts can explore options like supercharging or turbocharging to enhance the engine's output, providing a thrilling driving experience without compromising the car's agility.

The Subaru Legacy and Outback, though often overlooked in the performance tuning scene, offer unique opportunities for modifications. These models feature a more refined approach, combining everyday usability with the potential for performance upgrades. The turbocharged variants, particularly the Legacy GT, allow tuners to extract additional horsepower while maintaining practicality. Upgrades such as upgraded intercoolers, intake systems, and exhaust modifications can significantly enhance performance without detracting from the vehicles' versatility. This capability makes them appealing to those who desire a balance between performance and daily drivability.

Lastly, the Subaru Ascent, the brand's entry into the SUV segment, showcases Subaru's commitment to versatility and family-friendly design. While not traditionally associated with performance tuning, the Ascent's turbocharged engine and spacious interior offer a platform for those looking to enhance their family vehicle's capabilities. Modifications can focus on improving handling dynamics and power delivery, making it a compelling option for families who appreciate the thrill of driving. With a growing aftermarket presence, tuners can explore options to elevate the Ascent's performance while retaining its practicality.

In summary, Subaru's diverse lineup presents automotive modders and tuners with a wealth of opportunities for performance enhancements. Each model, from the rally-bred WRX to the versatile Outback and the agile BRZ, offers unique characteristics that can be tailored to individual preferences and driving styles. By understanding the potential of these vehicles, enthusiasts can embark on a journey of performance tuning that not only amplifies their Subaru's capabilities but also deepens their connection to the brand and the driving experience.

Understanding the Subaru Engine Architecture

Subaru has carved a unique niche in the automotive landscape with its distinctive engine architecture, which serves as a foundation for the brand's performance and reliability. At the heart of this architecture is the Subaru Boxer engine, characterized by its horizontally opposed cylinders. This design not only contributes to a lower center of gravity, enhancing stability and handling, but also reduces vibrations typically associated with conventional inline or V-shaped engines. Understanding the intricacies of this architecture is essential for automotive modders and tuners seeking to maximize performance and efficiency in their Subaru builds.

The Boxer engine's configuration allows for a more compact design, which is particularly advantageous in terms of weight distribution and space utilization within the chassis. This layout leads to a more balanced vehicle, resulting in improved cornering capabilities and overall driving dynamics. Additionally, the engine's flat design enables a more direct transfer of power to the drivetrain, reducing the need for complex and heavy components. As tuners delve into performance upgrades, recognizing the benefits of this architecture can guide their choices in modifications, such as exhaust systems and intake manifolds, that align with the engine's unique characteristics.

Subaru engines are renowned for their robust construction and inherent durability, making them suitable for performance enhancements. The use of forged internals in many models ensures that the engine can withstand increased power outputs without compromising reliability. Understanding the material choices and engineering principles behind the engine's design is crucial for tuners looking to push the limits of performance. For example, when considering turbocharger upgrades, it's vital to evaluate how increased boost pressure will affect engine longevity and to choose components that complement the Boxer architecture rather than simply overpower it.

Another critical aspect of Subaru engine architecture is the unique firing order and how it influences engine performance. The 180-degree crankshaft configuration results in a distinct exhaust note and power delivery, which can be particularly appealing to enthusiasts. This firing order also impacts tuning strategies, as it affects turbo spool characteristics and throttle response. Automotive modders need to be aware of these nuances when selecting tuning maps or adjusting fuel delivery systems. A well-rounded understanding of the engine's behavior will enable tuners to optimize their setups for both daily driving and track performance.

Finally, embracing the Subaru engine architecture means recognizing the potential for compatibility with aftermarket performance parts. Many manufacturers design components specifically for the Boxer engine, taking into account its unique characteristics and power band. From high-flow fuel injectors to precision-engineered turbochargers, the aftermarket offers a wealth of options that can enhance performance while maintaining reliability. Tuners should prioritize products that have been tested with Subaru's architecture to ensure compatibility and effectiveness, thereby maximizing the potential of their builds. By understanding the intricacies of the Subaru engine architecture, automotive enthusiasts can embark on a journey of performance tuning that is both rewarding and exhilarating.

Chapter 2: Setting Goals for Your Build

Defining Performance Objectives

Defining performance objectives is a crucial step for automotive modders and tuners, particularly in the realm of Subaru performance tuning. Establishing clear and measurable goals allows enthusiasts to focus their efforts effectively, ensuring that modifications align with desired outcomes. This process begins with understanding the specific aspects of performance that are important to the individual or team. Whether the objective is to increase horsepower, improve torque, enhance fuel efficiency, or optimize handling, each goal requires a tailored approach.

To effectively define performance objectives, it is essential to consider the intended use of the vehicle. For some, the primary focus may be on street performance, where a balance between daily drivability and spirited driving is key. Others might prioritize track performance, seeking maximum output and responsiveness at higher RPMs. Additionally, some enthusiasts may wish to explore off-road capabilities, necessitating a different set of modifications that enhance durability and traction. By identifying the primary purpose of the vehicle, tuners can create performance objectives that cater to specific driving conditions.

Another vital consideration in defining performance objectives is the budget and resources available. Performance tuning can quickly escalate in cost, especially when considering high-quality parts and professional installation. Setting realistic financial parameters helps prevent overspending and encourages strategic planning. This includes prioritizing modifications that offer the best return on investment, whether through increased performance, reliability, or resale value. Documenting the budget alongside performance goals allows for a systematic approach to building the vehicle, ensuring that each modification contributes to the overarching vision.

Measurable objectives play a significant role in the performance tuning process. Establishing quantifiable benchmarks, such as target horsepower numbers, lap times, or quarter-mile drag times, provides a clear framework for evaluating the success of modifications. This metric-driven approach also facilitates ongoing assessment, allowing tuners to adjust their strategies as needed. For instance, if a specific modification does not yield the expected results, re-evaluating the objectives and making informed decisions about future changes can lead to better outcomes. Utilizing tools like dyno runs or track day results offers concrete data to inform these evaluations.

Finally, it's essential to remain adaptable in the pursuit of performance objectives. The automotive landscape is constantly evolving, with new technologies and tuning methods emerging regularly. Staying informed about industry trends and advancements can lead to new opportunities for achieving performance goals. Additionally, the dynamic nature of vehicle performance means that objectives may need to evolve over time based on experiences and new insights. By fostering a mindset of continuous learning and flexibility, Subaru enthusiasts can ensure that their tuning journey remains exciting and rewarding, ultimately resulting in a vehicle that truly reflects their aspirations.

Street vs. Track Performance

In the world of automotive performance, particularly among Subaru enthusiasts, the debate between street and track tuning is a pivotal consideration. Each avenue presents unique challenges and benefits, making the choice largely dependent on individual goals and preferences. Street performance tuning often prioritizes drivability and comfort, ensuring that a vehicle remains practical for daily use. This involves finding the right balance of power, suspension setup, and tire selection to handle variable conditions while still providing an exhilarating driving experience. On the other hand, track performance tuning focuses on maximizing a vehicle's capabilities in a controlled environment, emphasizing speed, handling, and agility. Understanding these differences is essential for anyone looking to optimize their Subaru for a specific purpose.

When tuning for street performance, the primary goal is often to enhance the vehicle's responsiveness without sacrificing everyday usability. This can involve modifications such as upgrading the air intake system, remapping the ECU, or installing a more efficient exhaust system. These changes can significantly increase horsepower and torque while maintaining a smooth power delivery that is manageable in urban traffic. Additionally, suspension modifications may include adjustable coilovers or sway bars, providing a sportier feel without compromising ride comfort. The key is to create a setup that allows for spirited driving while remaining compliant enough for daily commutes.

Conversely, track performance tuning is laser-focused on achieving the highest possible levels of speed and handling. This typically involves more aggressive modifications, including weight reduction, enhanced brake systems, and specialized tires designed for optimal grip. Engine tuning for track use often pushes the limits of performance, utilizing advanced techniques such as larger turbochargers, upgraded fuel systems, and high-performance intercoolers. It's not uncommon for track-focused Subarus to sacrifice some degree of comfort and everyday drivability in favor of raw performance. This approach requires careful consideration of how modifications will impact reliability and maintenance, as track conditions can be unforgiving.

One of the most significant aspects of the street versus track performance debate is tire selection. Street performance tires are designed for versatility, offering a balance between grip and comfort for varied driving conditions. These tires need to withstand wet and dry conditions while providing adequate feedback and longevity. In contrast, track tires are engineered for maximum grip and heat resistance, often sacrificing tread life and comfort in the process. The choice of tires can drastically affect handling characteristics, acceleration, and braking distances, making it a crucial component of any performance build.

Ultimately, the decision between street and track performance tuning hinges on the driver's priorities. While some enthusiasts may seek the thrill of high-speed laps on a closed circuit, others may prefer the joy of a well-tuned daily driver that can tackle winding roads and urban environments with equal flair. Understanding the fundamental differences in tuning philosophy allows Subaru modders and tuners to tailor their projects to fit their lifestyle and driving aspirations. Whether the goal is to conquer the local track or enjoy spirited drives through scenic routes, mastering the art of performance tuning means striking the right balance between power, handling, and practicality.

Budgeting for Modifications

Budgeting for modifications is a crucial step for any automotive enthusiast, particularly for those diving into the world of Subaru performance tuning. With a seemingly endless array of aftermarket parts, upgrades, and tuning options available, it is easy to get carried away and overspend. Establishing a budget not only helps in managing expenses but also ensures that each modification aligns with your overall performance goals. By carefully planning your financial approach, you can maximize the potential of your Subaru while avoiding the pitfalls of impulsive spending.

A well-thought-out budget begins with setting clear objectives for your vehicle. Determine what kind of performance enhancements you desire—whether it's increased horsepower, improved handling, or enhanced aesthetics. Identifying your primary goals allows you to prioritize which modifications will deliver the most value for your investment. For instance, if your focus is on racing, you might allocate more funds towards suspension upgrades and lightweight components. Conversely, if daily drivability is your concern, investing in a quality engine tune and exhaust system may take precedence.

Next, research the costs associated with your desired modifications. Each component comes with its own price tag, which can vary significantly between brands and performance levels. Additionally, consider installation costs if you're not planning to undertake the work yourself. Many enthusiasts underestimate the expenses involved in professional installation, which can add a considerable amount to the overall budget. Create a detailed list of potential parts and services, noting both the prices and any shipping or labor fees. This will give you a clearer picture of the total investment required.

Another important aspect of budgeting is planning for unexpected expenses. When modifying performance vehicles, unforeseen issues can arise, such as needing additional parts or encountering problems during installation. It's wise to set aside a contingency fund—typically around 10-20% of your total budget—to cover these surprises. This not only helps to cushion the financial blow but also prevents you from being forced to abandon a project mid-way due to a lack of funds. Preparation can be the difference between a successful modification journey and a frustrating experience.

Finally, keep in mind that performance tuning is an ongoing process. As you complete one modification, you may discover new areas for improvement or additional upgrades that pique your interest. Therefore, it is prudent to maintain a flexible budget that allows for future modifications. Regularly review your budget and adjust it based on your experiences, performance goals, and new trends in the Subaru tuning community. By maintaining a proactive approach to budgeting, you will find yourself better equipped to achieve your dream build while enjoying the journey of performance tuning.

Chapter 3: Basic Modifications

Intake Systems

Intake systems play a crucial role in the performance tuning of Subaru vehicles, serving as the gateway for air to enter the engine. The efficiency of this system directly influences the engine's ability to produce power, making it a focal point for modders and tuners looking to maximize performance. The fundamental principle behind an intake system is to facilitate the smooth flow of air into the combustion chamber, optimizing the air-fuel mixture for combustion. This is particularly important in turbocharged Subarus, where enhanced airflow can lead to significant power gains and improved throttle response.

When considering modifications to the intake system, one must first understand the various components involved. These typically include the air filter, intake piping, and the intake manifold. Each of these elements contributes to the overall effectiveness of the system. Aftermarket air filters, for instance, are designed to provide better airflow than stock options while still offering adequate filtration. Upgrading to a high-flow air filter can lead to a noticeable increase in horsepower, especially when paired with larger diameter intake piping that reduces restrictions and improves airflow velocity.

Another critical aspect of intake systems is the role of the intake manifold. Different designs can greatly affect how air is distributed to each cylinder. For performance tuning, tuners may opt for a manifold that enhances airflow characteristics, allowing for better volumetric efficiency at higher RPMs. Additionally, some aftermarket manifolds come with features like integrated nitrous ports or throttle body spacers, which can further enhance performance potential. Choosing the right intake manifold can be a game changer for those looking to push their Subaru to new performance heights.

Heat management is another consideration when optimizing intake systems. Hot air is less dense than cold air, which can lead to a decrease in engine performance. Therefore, many tuners invest in heat shields or even insulated intake piping to keep the incoming air cooler. Cold air intakes are particularly popular as they draw air from outside the engine bay, where temperatures are lower. This simple yet effective modification can lead to significant improvements in performance, particularly in high-stress driving conditions or during track days.

Finally, it's essential to remember that any modifications to the intake system should be complemented by appropriate tuning. Adjustments to the engine management system are necessary to ensure that the new air-fuel ratios are optimized for the increased airflow. This is where the expertise of a skilled tuner comes into play, as they can recalibrate parameters to maximize performance gains while maintaining engine reliability. Whether you're aiming for a street setup or a track-focused build, investing time and resources into perfecting your Subaru's intake system is a vital step on the journey to mastering performance tuning.

Exhaust Upgrades

Exhaust upgrades are a vital aspect of performance tuning for Subaru enthusiasts. The factory exhaust systems are often restrictive, designed primarily for noise reduction and emissions compliance rather than maximizing engine output. Upgrading the exhaust system can significantly enhance the vehicle's performance by improving exhaust flow, reducing back pressure, and providing a more aggressive sound that aligns with the sporty character of Subaru models. Whether you're aiming for a modest increase in power or preparing for more extensive engine modifications, selecting the right exhaust components is crucial.

One of the most common upgrades is the installation of a high-performance cat-back exhaust system. This system includes everything from the catalytic converter back, replacing the stock muffler and piping with larger diameter components that facilitate better airflow. Brands like Cobb, HKS, and Borla offer systems specifically engineered for various Subaru models, ensuring compatibility and optimal performance gains. These upgrades can lead to noticeable improvements in horsepower and torque, as well as a more pronounced and robust exhaust note that many Subaru enthusiasts crave.

Another essential component in exhaust upgrades is the downpipe. Upgrading to a high-flow downpipe, particularly one that eliminates the factory catalytic converter or uses a high-performance cat, can dramatically reduce exhaust restrictions. A well-designed downpipe allows for quicker turbo spool and improved throttle response, which is especially beneficial for turbocharged Subarus like the WRX and STI. However, it's crucial to consider local emissions regulations when opting for a downpipe that removes the catalytic converter, as this can affect legality and vehicle inspections.

Beyond cat-back systems and downpipes, headers are another area where performance can be significantly enhanced. Upgrading to aftermarket headers can improve exhaust flow from the engine, reducing back pressure and allowing for more efficient combustion. This can lead to an increase in power and torque, particularly in naturally aspirated Subaru models. When selecting headers, tuners should consider the material (stainless steel is preferable for durability), design (equal-length vs. unequal-length), and compatibility with other exhaust components to ensure a seamless integration into the overall exhaust setup.

Finally, it's essential to pair exhaust upgrades with proper tuning to maximize their effectiveness. An upgraded exhaust system changes the dynamics of the engine's air-fuel mixture, necessitating adjustments to the vehicle's tuning parameters. Using a performance tuner or ECU reflash can help optimize the engine's performance and ensure that the vehicle runs smoothly with the new exhaust setup. By investing in a quality exhaust upgrade and proper tuning, Subaru enthusiasts can unlock significant performance potential, taking their driving experience to exhilarating new heights.

ECU Reflashing and Tuning

ECU reflashing and tuning is a vital aspect of enhancing Subaru performance, allowing enthusiasts and modders to unlock the full potential of their vehicles. The Engine Control Unit (ECU) is essentially the brain of the vehicle, controlling various engine parameters such as fuel delivery, ignition timing, and boost pressure. By reflashing the ECU, tuners can modify these parameters to optimize performance, improve drivability, and increase fuel efficiency. This process is especially important for those who have upgraded their Subaru with aftermarket parts, as stock tuning may not account for the changes made through modifications like larger turbochargers, enhanced intercoolers, or high-flow exhaust systems.

The process of ECU reflashing involves overwriting the factory programming with a custom tune designed to suit the specific needs of the modified engine. This is typically achieved using specialized software and hardware tools, which allow tuners to adjust numerous variables that influence engine performance. For Subaru owners, the ability to fine-tune parameters such as air-fuel ratios, boost levels, and rev limits can lead to significant gains in horsepower and torque. Moreover, by optimizing these settings, tuners can ensure that their vehicles run smoothly under various conditions, reducing the risk of engine knock or other performance-related issues.

One of the most critical aspects of successful ECU tuning is understanding the balance between performance and reliability. While it can be tempting to push the limits of power output, a well-rounded tune should also consider the longevity of the engine and its components. For Subaru engines, which are often subjected to high levels of stress from modifications, maintaining a safe operating range is essential. This involves using data from dyno runs and real-world driving to identify the ideal settings that maximize performance without compromising engine safety.

In addition to performance gains, ECU reflashing can also enhance the overall driving experience. Many Subaru enthusiasts find that a properly tuned ECU can lead to improved throttle response, smoother power delivery, and a more engaging driving feel. Tuning can also help eliminate common issues associated with modified vehicles, such as rough idling or stalling, by ensuring that the engine operates efficiently across all RPM ranges. This results in a more enjoyable experience on the road or track, where every ounce of performance and responsiveness counts.

Finally, it's important for automotive modders and Subaru fiends to stay informed about the latest developments in ECU tuning technology. As manufacturers release new models and updates, the tuning landscape continually evolves, offering new opportunities for performance enhancement. Engaging with the community through forums, social media, and tuning events can provide valuable insights and tips for both novice and experienced tuners. By embracing the art of ECU reflashing and tuning, Subaru enthusiasts can not only elevate their vehicles' performance but also deepen their understanding of automotive engineering, ultimately leading to a more fulfilling modding journey.

Chapter 4: Advanced Engine Modifications

Turbocharger Upgrades

Turbocharger upgrades represent one of the most effective methods to increase the performance of Subaru engines, particularly in models such as the WRX and STI. As enthusiasts know, the factory-installed turbochargers are designed to balance performance and reliability, but they often leave room for improvement. Upgrading your turbocharger can unlock the potential of your Subaru's engine, allowing for higher horsepower and torque outputs. Understanding the intricacies of turbocharger options, installation procedures, and tuning requirements is essential for anyone looking to maximize their vehicle's performance.

When considering a turbocharger upgrade, it is crucial to evaluate the various options available in the market. Upgrades can range from larger factory-style replacements to high-performance aftermarket units designed for maximum efficiency and power. Options such as the Garret GTX series or the BorgWarner EFR line offer advanced technologies like ball-bearing centers and integrated wastegates, which can enhance spool time and reduce turbo lag. Additionally, hybrid turbos that utilize components from both OEM and aftermarket designs can provide a balanced approach to performance, allowing for a significant boost without sacrificing daily drivability.

Installation of a new turbocharger requires careful planning and execution. It's not merely a matter of swapping out the old unit; it involves a thorough understanding of the engine's supporting modifications. Upgraded fuel injectors, a high-flow fuel pump, and a performance intercooler are often necessary to handle the increased air flow and heat generated by a larger turbo. Furthermore, attention must be paid to the exhaust system, as a free-flowing setup will help reduce backpressure and improve overall efficiency. Properly aligning all components during installation is critical to avoid issues that can arise from improper fitment or inadequate support.

Tuning is an essential step following a turbocharger upgrade. While the new turbo can provide increased airflow, the engine management system must be recalibrated to take full advantage of the additional power. This process typically involves the use of a standalone engine management system or reprogramming the factory ECU. A well-executed tune ensures that the air-fuel ratio, ignition timing, and boost levels are optimized for the new setup, enhancing both performance and reliability. Without proper tuning, the risk of engine knock and other performance issues significantly increases, potentially leading to catastrophic engine failure.

Finally, it is vital for modders and tuners to understand the importance of regular maintenance and monitoring after a turbocharger upgrade. Keeping an eye on parameters such as boost levels, exhaust gas temperatures, and oil pressure can help in identifying potential issues before they escalate. Regularly scheduled maintenance, including oil changes with high-quality synthetic oils and periodic inspections of the turbo and related components, is necessary to ensure longevity and performance. By investing the time and resources into proper upgrades and maintenance, Subaru enthusiasts can enjoy the exhilarating power of a well-tuned, turbocharged engine while maintaining the reliability that these vehicles are known for.

Intercooler Systems

Intercooler systems play a pivotal role in the performance tuning landscape, particularly for Subaru enthusiasts who are keen on maximizing their vehicles' potential. An intercooler serves as a heat exchanger that cools the air entering the engine after it has been compressed by the turbocharger. This cooling effect is essential because cooler air is denser and contains more oxygen, allowing for a more efficient combustion process. For modders and tuners, understanding the intricacies of intercooler systems is crucial for achieving optimal performance and reliability in their Subaru builds.

There are two primary types of intercoolers used in performance applications: air-to-air and air-to-water. Air-to-air intercoolers are the most common in Subaru performance tuning. They utilize ambient air to cool the intake charge as it passes through the intercooler core. The design and placement of these intercoolers can significantly impact their efficiency; for example, a larger core with a well-optimized flow path will provide better cooling than a smaller, poorly positioned unit. Conversely, air-to-water intercoolers use a liquid coolant to absorb heat from the intake air, which can offer a more consistent cooling performance, especially in high-boost applications where ambient air temperatures might be less favorable.

The choice of intercooler can also influence other aspects of engine tuning. A well-designed intercooler can help reduce the risk of knock by lowering intake air temperatures, allowing tuners to run more aggressive timing maps and higher boost levels safely. However, installing a larger intercooler may necessitate additional modifications, such as upgraded piping and a more powerful cooling system, which can add complexity to the build. Tuning enthusiasts must consider these factors when selecting an intercooler to ensure a harmonious balance between performance and practicality.

In addition to the type and size of the intercooler, the quality of the intercooler core is paramount. High-performance intercoolers often feature bar-and-plate designs, which provide superior thermal efficiency compared to tube-and-fin designs typically found in stock units. This increased efficiency translates to better performance across various driving conditions, from daily commutes to track days. When selecting an intercooler, modders should prioritize high-quality materials and construction to ensure durability and effectiveness under the stresses of high-performance driving.

Finally, proper installation and maintenance of intercooler systems cannot be overlooked. Ensuring that all connections are secure and that there are no leaks is crucial for maintaining optimal performance. Regularly inspecting the system for debris and damage will help prevent any potential issues that could arise from wear and tear. By taking the time to understand and implement a well-designed intercooler system, Subaru enthusiasts can unlock the full potential of their vehicles, enhancing both performance and driving enjoyment in their pursuit of boosted dreams.

Fuel System Enhancements

In the realm of automotive performance tuning, particularly for Subaru enthusiasts, the fuel system is a critical component that often determines the limits of power and efficiency. As tuning progresses, stock fuel systems frequently become inadequate, unable to supply the necessary fuel to meet the demands of high-performance modifications. Upgrading the fuel system is not just about increasing horsepower; it's also about ensuring reliability and optimal performance under various driving conditions. This subchapter delves into the key enhancements that can be made to the fuel system of your Subaru, setting the foundation for a more robust and powerful driving experience.

One of the first considerations when enhancing the fuel system is the fuel pump. Stock fuel pumps are designed for factory specifications and may struggle to deliver the required fuel volume and pressure in high-performance applications. Upgrading to a high-flow fuel pump can significantly increase the fuel delivery capacity, ensuring that your engine receives the necessary fuel under heavy load. Aftermarket options typically provide higher flow rates and improved reliability, which are essential for supporting increased horsepower from turbocharging or other modifications. When selecting a fuel pump, it's crucial to consider the compatibility with your existing fuel system and the overall power goals of your build.

Next in line are fuel injectors. The injectors play a pivotal role in delivering the right amount of fuel to the engine in a precise manner. Stock injectors may become a bottleneck as power levels rise, as they may not be able to inject sufficient fuel at higher RPMs. Upgrading to larger, high-performance injectors can greatly enhance fuel atomization, resulting in better combustion efficiency and power output. It's important to match the injector size to the specific modifications and power goals of the engine, as oversized injectors can lead to poor drivability and fuel economy. Proper tuning after installation is essential to ensure that the air-fuel mixture remains optimal across the entire RPM range.

Incorporating a fuel management system can further enhance the performance of your Subaru's fuel system. These systems allow for precise control of fuel delivery, enabling tuners to adjust fuel maps according to the vehicle's modifications and driving conditions. Aftermarket fuel management solutions can provide real-time monitoring and adjustments, which is vital when running higher boost levels or varying fuel types. By utilizing a standalone engine management system or a piggyback tuner, enthusiasts can fine-tune fuel delivery for optimal performance, ensuring that the engine operates efficiently and safely. This flexibility can lead to significant gains in horsepower and torque, while also improving overall drivability.

Another critical aspect of fuel system enhancements is the fuel lines and delivery system. Stock fuel lines may not be designed to handle the increased pressures and fuel flow associated with high-performance setups. Upgrading to larger diameter, high-performance fuel lines can reduce restrictions and improve fuel flow to the engine. Additionally, using high-quality materials that can withstand higher temperatures and pressures is essential for long-term reliability. It's also wise to consider the routing of fuel lines to avoid heat soak and potential vapor lock issues, which can adversely affect performance, especially in high-stress situations like racing or spirited driving.

Finally, don't overlook the importance of proper fuel quality and filtration in your performance tuning journey. Using high-octane fuel can prevent knocking and ensure that the engine runs smoothly under boost. Additionally, upgrading the fuel filter to a high-performance option can help prevent contaminants from entering the fuel system, which is vital for maintaining the health of the fuel pump and injectors. Regular maintenance and monitoring of the fuel system will not only prolong the life of these components but also ensure that your Subaru delivers the performance you desire. By implementing these fuel system enhancements, you lay the groundwork for a powerful and reliable performance vehicle that can handle the demands of tuning and modification.

Chapter 5: Suspension and Handling

Upgrading Shocks and Struts

Upgrading shocks and struts is a crucial modification for Subaru enthusiasts looking to enhance their vehicle's performance, handling, and overall driving experience. While many may focus on engine tuning and power upgrades, the suspension system plays an equally important role in how a car responds to the road and behaves during spirited driving or competitive events. By investing in high-quality shocks and struts, you can significantly improve your vehicle's stability, cornering abilities, and ride comfort, allowing you to fully enjoy the rewards of your hard work and investment.

The stock suspension components on most Subaru models are designed with a balance of comfort and performance in mind, which may not be suitable for those who push their vehicles to the limit. Upgrading to performance shocks and struts can provide a much-needed boost in responsiveness and control. These upgraded components often feature adjustable settings that allow owners to fine-tune their ride to match personal preferences and driving conditions. This adaptability is invaluable for those who participate in various motorsport events or simply enjoy the thrill of driving their modified Subaru on twisty backroads.

When selecting shocks and struts, it's essential to consider factors such as the type of driving you plan to do, your vehicle's current modifications, and your budget. There are numerous options available, ranging from coilovers to performance shock absorbers. Coilover systems offer the added benefit of adjustable ride height, allowing you to dial in the perfect stance while optimizing handling characteristics. On the other hand, performance shock absorbers can be paired with stock springs or aftermarket springs for a more straightforward upgrade that still yields significant improvements in performance.

Installation of upgraded shocks and struts can vary in complexity depending on your chosen components and your mechanical skill level. While some enthusiasts may opt to tackle the installation themselves, others may prefer to seek the expertise of a professional mechanic or suspension specialist. Regardless of your approach, it's crucial to follow the manufacturer's instructions and ensure that all components are installed correctly. Proper alignment after installation is also essential to prevent uneven tire wear and maximize the benefits of your new suspension setup.

Finally, remember that upgrading shocks and struts is just one part of the overall performance tuning process. For the best results, consider complementing your suspension upgrades with other modifications, such as sway bars, bushings, and chassis bracing. These enhancements work together to create a more cohesive and responsive driving experience. By prioritizing suspension upgrades alongside engine modifications, you can achieve a well-rounded performance vehicle that excels both on the track and the road, ultimately bringing your boosted dreams to life.

Sway Bars and Chassis Stiffening

Sway bars and chassis stiffening are critical components in enhancing the performance of Subaru vehicles, especially for those who are passionate about performance tuning and automotive modifications. These elements play a significant role in managing the dynamics of your car during cornering and high-speed maneuvers. By understanding how sway bars function and the importance of a stiff chassis, modders can unlock the full potential of their Subaru, transforming it into a finely tuned machine capable of handling the rigors of both the street and the track.

Sway bars, or anti-roll bars, are designed to reduce body roll during cornering. They connect the left and right wheels of a vehicle, allowing them to work together to maintain stability. When a Subaru takes a corner, the sway bar helps distribute the load more evenly across the tires, which can significantly improve grip. For automotive tuners, upgrading to a thicker sway bar can result in a more responsive handling experience. The increased stiffness reduces the amount of body roll, allowing for a flatter cornering stance. This upgrade not only enhances performance but also provides a more engaging driving experience, making it a popular choice among Subaru enthusiasts.

In addition to sway bars, chassis stiffening is a crucial modification for those seeking to improve their Subaru's performance. A stiff chassis reduces flex during aggressive driving, ensuring that the suspension geometry remains constant. This consistency is vital for maintaining optimal tire contact with the road, which translates to improved handling and stability. Various methods can be employed to stiffen the chassis, including the installation of strut braces, chassis braces, and even a full roll cage for serious track enthusiasts. Each of these modifications contributes to a more rigid structure, allowing the suspension to work more effectively and providing better feedback to the driver.

It's important to strike a balance when modifying sway bars and chassis stiffness. While a stiffer setup can improve handling, overly stiff components can lead to a harsh ride quality, which may detract from everyday drivability. For Subaru modders, it's essential to consider the intended use of the vehicle. A car set up strictly for track use may benefit from increased stiffness, while a daily driver should prioritize a balance that retains comfort without sacrificing too much performance. Testing different setups and adjustments can help find that sweet spot, allowing for a car that performs well under various conditions.

Ultimately, the combination of upgraded sway bars and reinforced chassis creates a more dynamic driving experience for Subaru enthusiasts. These modifications not only enhance the performance of the vehicle but also instill a sense of confidence in the driver. As automotive modders continue to explore the limits of their Subarus, understanding the relationship between sway bars and chassis stiffness becomes essential. By making informed choices and tuning these aspects effectively, enthusiasts can achieve a level of performance that transforms their Subaru into a powerhouse on both the street and the track, living up to the dreams of every automotive tuner.

Tire and Wheel Selection

Tire and wheel selection is a critical aspect of performance tuning that can significantly affect your Subaru's handling, acceleration, and overall driving experience. For automotive modders and tuners, the right combination of tires and wheels not only enhances aesthetic appeal but also optimizes performance based on specific driving conditions and preferences. When considering upgrades, it's essential to understand how these components interact with the vehicle's suspension, weight distribution, and power output.

When selecting wheels, size and weight are primary considerations. Larger wheels can improve cornering stability and provide a more aggressive look, but they can also introduce a trade-off in terms of ride comfort and acceleration. Additionally, wheel weight plays a crucial role in performance; lighter wheels reduce unsprung mass, improving suspension response and overall agility. As a rule of thumb, aim for wheels that maintain or improve upon your Subaru's factory specifications while also accommodating your tire choices and driving style.

Tires are equally vital in maximizing your vehicle's performance. The available options can be overwhelming, with categories ranging from all-season to ultra-high-performance tires. When tuning for performance, focus on tires that align with your driving objectives, whether that be daily driving, track days, or off-road adventures. Factors such as tread pattern, rubber compound, and temperature range all influence traction, handling, and wear characteristics. Understanding the specific needs of your Subaru, based on its tuning and intended use, will guide you in selecting the optimal tire type.

Another crucial element to consider is the tire's aspect ratio, which affects its profile and handling characteristics. A lower aspect ratio generally provides better stability and cornering performance due to a wider contact patch, which enhances grip during aggressive maneuvers. However, this can come at the cost of ride comfort, as lower-profile tires tend to transmit more road imperfections to the cabin. Finding a balance that suits your driving style and comfort preferences is essential, especially for a vehicle that may be used for daily driving alongside performance-focused applications.

Finally, the fitment of your selected wheels and tires must be taken into account to avoid issues such as rubbing or clearance problems. Ensure that your choices comply with your Subaru's suspension setup and any modifications that have been made. Utilizing tools like a tire size calculator can help you visualize the differences in diameter and width, allowing for precise adjustments. By carefully considering the interplay between tire and wheel selection, you can achieve not only enhanced performance but also a tailored driving experience that embodies the spirit of Subaru tuning.

Chapter 6: Transmission and Drivetrain Modifications

Upgrading Clutches and Flywheels

Upgrading clutches and flywheels is an essential consideration for automotive modders and tuners looking to enhance the performance of their Subaru vehicles. As you increase power output through modifications such as turbo upgrades and engine tuning, the factory components may struggle to handle the enhanced torque and horsepower. This subchapter will delve into the critical aspects of selecting the right clutch and flywheel to ensure your Subaru can effectively transfer power to the wheels without compromising performance or drivability.

The clutch is the heart of the drivetrain, responsible for engaging and disengaging the engine from the transmission. Upgrading to a high-performance clutch can provide faster engagement, improved pedal feel, and increased torque capacity. When choosing a clutch, it's essential to consider the type of driving you'll be doing. Options range from organic disc clutches, which offer smooth engagement and are suitable for daily driving, to ceramic and puck-style clutches, which can handle more abuse but may sacrifice drivability and smoothness. Each type has its advantages and disadvantages, making it crucial to match the clutch choice with your performance goals and driving style.

Equally important is the flywheel, which plays a vital role in engine performance. A lightweight flywheel can improve throttle response and acceleration by reducing rotational mass, allowing the engine to rev more freely. However, lighter flywheels can also lead to a loss of low-end torque, which may affect drivability in daily driving scenarios. Depending on your tuning objectives, you may opt for a lightweight aluminum flywheel or a dual-mass flywheel for a more balanced approach that retains some of the factory characteristics while still offering performance benefits. Understanding these trade-offs will help you make an informed decision that aligns with your overall performance strategy.

Installation of upgraded clutches and flywheels should not be taken lightly, as improper installation can lead to premature failure or even damage to other drivetrain components. It is recommended to utilize professional services unless you have extensive experience with Subaru transmissions. Additionally, consider the importance of breaking in new components properly to ensure longevity and optimal performance. Many manufacturers provide specific guidelines for breaking in their clutches, which typically involve a period of moderate driving to allow the materials to settle and bond effectively.

In conclusion, upgrading clutches and flywheels is a critical step in maximizing the performance potential of your Subaru. By carefully considering your driving style and performance objectives, you can make informed choices that enhance your vehicle's capabilities. Balancing factors such as engagement feel, torque capacity, and weight will ultimately lead to a more rewarding driving experience. As you embark on this journey of performance tuning and engine modifications, remember that the right clutch and flywheel combination can make all the difference in unleashing the full potential of your Subaru.

Differential Enhancements

Differential enhancements play a critical role in maximizing the performance of Subaru vehicles, particularly for those who engage in spirited driving or competitive motorsports. The differential is responsible for distributing power from the engine to the wheels, and its design can significantly influence traction, handling, and overall driving dynamics. In the realm of performance tuning, understanding the various types of differential enhancements available can provide enthusiasts with the tools necessary to optimize their Subaru's performance for both street and track conditions.

One of the most common enhancements is the installation of a limited-slip differential (LSD). Unlike an open differential that allows one wheel to spin faster than the other, a limited-slip differential helps to transfer power to the wheel with more traction. This is particularly beneficial for Subaru models, which often feature all-wheel drive systems. There are several types of LSDs available, including clutch-type, helical, and viscous. Each has its advantages and is suited to different driving styles and conditions. For instance, a clutch-type LSD offers adjustable settings for performance tuning, while a helical LSD provides a more seamless engagement, making it ideal for daily drivers.

Another critical aspect of differential enhancements is the gear ratio. Changing the differential gear ratio can significantly alter the vehicle's acceleration characteristics and top speed. Lowering the gear ratio (numerically higher) can improve acceleration by allowing the engine to reach its power band more quickly, which is advantageous for track use. Conversely, a higher gear ratio can enhance fuel efficiency and top-end performance, making it suitable for highway driving. Understanding the trade-offs involved in gear ratio changes is essential for tuners looking to tailor their Subaru's performance to specific uses.

For those pushing the limits of their Subaru's capabilities, considering upgraded differential mounts is also important. Stock mounts can often be soft and allow for excessive movement under load, leading to inconsistent power delivery and compromised handling. Upgraded mounts made from polyurethane or other high-stiffness materials can minimize differential movement, ensuring that power is effectively transferred to the wheels. This added stability not only improves traction but also enhances cornering performance, which is crucial in competitive scenarios.

In addition to mechanical enhancements, tuning the vehicle's electronic systems can also yield benefits for differential performance. Many modern Subaru models come equipped with advanced traction and stability control systems that can be calibrated to work harmoniously with upgraded differentials. By fine-tuning these electronic systems, tuners can optimize the vehicle's response during acceleration, braking, and cornering. This holistic approach to differential enhancements ensures that both mechanical and electronic components work together seamlessly, resulting in a Subaru that is not only powerful but also agile and responsive on the road or track.

Short Shifters and Gear Ratios

Short shifters and gear ratios play a significant role in enhancing the performance and driving experience of Subaru vehicles, particularly for those who are passionate about automotive modification and tuning. A short shifter is a crucial component that allows for quicker gear changes, reducing the distance the shifter must travel to engage each gear. This not only improves the responsiveness of the transmission but also contributes to a sportier feel, giving drivers a more connected experience with their vehicle. For Subaru enthusiasts looking to maximize their performance, understanding the implications of short shifters and gear ratios is essential.

When considering a short shifter, it's important to evaluate the design and quality of the product. A well-engineered short shifter can significantly decrease the height of the gear lever and the throw distance, making gear changes more efficient. This reduction in throw not only allows for faster shifts but also helps maintain driver focus on the road and track, rather than the mechanics of shifting. Brands that specialize in performance parts often offer adjustable short shifters, allowing modders to customize the throw length to their preference, thus optimizing the driving experience.

In conjunction with short shifters, gear ratios are another critical factor in achieving optimal performance. The gear ratio determines how many times the output shaft turns for every turn of the input shaft, affecting acceleration and top speed. A lower gear ratio, for instance, can provide quicker acceleration, making it ideal for track racing or spirited driving. Conversely, a higher gear ratio offers better fuel efficiency and a smoother ride at highway speeds. Subaru owners should carefully consider the intended use of their vehicle when selecting gear ratios to ensure they strike a balance between performance and practicality.

Installing a short shifter often involves more than simply replacing the shifter itself; it may require adjustments to the linkage and potentially modifications to the transmission tunnel. This makes it essential for automotive modders to have a good understanding of their vehicle's layout and the specific parts involved in the installation process. Furthermore, integrating a short shifter with aftermarket gear ratios can amplify performance gains, allowing for a truly personalized driving experience. Compatibility between the shifter and the transmission is crucial, as mismatched components can lead to shifting issues or even damage.

Ultimately, the combination of short shifters and optimized gear ratios can radically transform the dynamics of a Subaru. For automotive tuners seeking to enhance their vehicle's performance, investing in quality short shifters and carefully chosen gear ratios can yield impressive results. By understanding how these components interact and affect driving behavior, Subaru enthusiasts can make informed decisions that align with their performance goals, ensuring that every shift is as exhilarating as the last. Whether on the track or the open road, mastering this aspect of performance tuning will elevate the driving experience to new heights.

Chapter 7: Tuning for Performance

Understanding Dyno Testing

Dyno testing is an essential tool for automotive modders and tuners, particularly for those passionate about Subaru performance. At its core, a dynamometer, or dyno, is a device used to measure force, torque, and power output from an engine or vehicle. Understanding the principles behind dyno testing can significantly enhance the tuning process, allowing enthusiasts to make informed decisions about modifications and improvements. For Subaru owners, this knowledge is key to unlocking the full potential of their vehicles and achieving the desired performance outcomes.

There are two primary types of dynamometers used in automotive testing: the engine dynamometer and the chassis dynamometer. An engine dyno measures the power output of the engine itself in a controlled environment, detached from the vehicle. This method allows tuners to analyze engine performance without the influence of other components. Conversely, a chassis dyno assesses the power output at the wheels, accounting for all drivetrain losses. This is particularly important for Subaru enthusiasts, as it provides a more realistic representation of the vehicle's performance under real-world conditions. Understanding the differences between these two testing methods is crucial for tuners aiming to optimize their setups.

The process of dyno testing involves several key steps that can provide invaluable data for performance tuning. First, the vehicle is secured on the dyno, and its initial baseline performance is recorded. This step is essential as it establishes a reference point against which future modifications can be measured. After the baseline run, various tuning adjustments—such as changes to the fuel mapping, boost levels, or timing—can be made, followed by additional runs to gauge the impact of these modifications. This iterative process allows tuners to fine-tune their setups, ensuring that each adjustment contributes positively to performance.

Data collected from dyno testing is not merely numbers; it tells a story about the engine's behavior under different conditions. For Subaru tuners, this data can reveal crucial insights, such as the engine's torque curve, horsepower peak, and responsiveness at various RPM levels. Understanding these metrics helps modders identify areas for improvement and make strategic choices regarding future modifications. For instance, tuners can analyze whether increasing boost pressure results in a linear increase in power or if it leads to diminishing returns, allowing them to optimize performance without risking engine reliability.

In conclusion, understanding dyno testing is vital for any Subaru enthusiast looking to maximize their vehicle's performance. By grasping the principles behind dynamometers, the nuances of testing methods, and the interpretation of collected data, tuners can make informed decisions that lead to significant gains in power and efficiency. As the tuning community continues to evolve, embracing the knowledge gained from dyno testing will empower modders to push the boundaries of performance, turning their boosted dreams into reality.

Fuel Mapping and Timing Adjustments

Fuel mapping and timing adjustments are crucial elements in optimizing the performance of a Subaru engine. These modifications are not merely about increasing power; they serve to enhance efficiency, improve throttle response, and ensure that the engine runs smoothly across a wide range of conditions. For automotive modders and tuners, understanding the intricacies of fuel mapping and timing is essential for unlocking the full potential of their Subaru vehicles.

Fuel mapping refers to the process of adjusting the air-fuel mixture that enters the engine. This is accomplished through the engine control unit (ECU), which relies on various sensors to determine the optimal amount of fuel needed at any given moment. Different modifications, such as upgraded turbochargers, intercoolers, or exhaust systems, can alter the air intake characteristics, necessitating changes in the fuel map. A well-tuned fuel map ensures that the engine receives the right amount of fuel for the given air conditions, maximizing power output while minimizing the risk of detonation or engine knock.

Timing adjustments, on the other hand, involve altering the ignition timing to ensure that the air-fuel mixture ignites at the optimal moment. The timing of ignition directly impacts engine performance, efficiency, and emissions. Advancing the timing can lead to increased power but also raises the risk of knock if not carefully managed. Conversely, retarding the timing can help prevent knock but may lead to a loss of power. For Subaru enthusiasts, understanding how to balance these adjustments is critical, especially when targeting specific power goals or when dealing with higher levels of boost pressure.

When tuning a Subaru, it is essential to utilize a wideband O2 sensor to monitor the air-fuel ratio (AFR) in real-time. This data allows tuners to make informed adjustments to the fuel map, ensuring that the engine operates within safe AFR limits, typically between 11.5:1 and 12.5:1 for a performance setup. Additionally, utilizing a dyno for tuning sessions can provide valuable feedback on how fuel mapping and timing adjustments affect overall power output and torque curves. This empirical data aids in refining the tune to achieve the best possible performance while maintaining engine reliability.

Ultimately, mastering fuel mapping and timing adjustments is a journey that requires both knowledge and experience. The Subaru community is rich with resources, including forums, tuning software, and dedicated shops, where enthusiasts can share their insights. For automotive modders and tuners looking to elevate their Subaru performance, investing time in learning about these critical aspects of tuning will not only enhance their vehicle's capabilities but also deepen their understanding of automotive engineering and performance dynamics. With the right approach, the dreams of boosted performance can become a reality.

Troubleshooting Common Issues

Troubleshooting common issues in Subaru performance tuning is a crucial skill for any automotive modder or tuner looking to maximize their vehicle's potential. Whether you're a seasoned enthusiast or just starting on your tuning journey, understanding how to diagnose and resolve common problems can save you time, money, and frustration. This subchapter will highlight some prevalent issues you may encounter, offer practical solutions, and provide tips for maintaining your Subaru's performance integrity.

One of the most frequent issues faced by Subaru tuners is improper air-fuel mixture, which can lead to performance inefficiencies and potential engine damage. Symptoms of this problem may include rough idling, poor acceleration, or unexpected engine stalling. To troubleshoot, first check your air intake and fuel system components, including the mass airflow sensor (MAF) and fuel injectors. A clean MAF sensor and properly functioning injectors are essential for maintaining the correct air-fuel ratio. Additionally, consider re-evaluating your tuning software settings, as incorrect maps can also lead to mixture discrepancies.

Another common issue involves ignition timing. If your Subaru experiences knocking or pinging, particularly under heavy acceleration, it may be a sign of timing problems. Begin troubleshooting by examining your ignition system components, including spark plugs, ignition coils, and wiring. Replacing worn spark plugs can sometimes resolve pre-ignition issues, but it's equally important to ensure your tuning software reflects the optimal timing adjustments for your modifications. Utilizing a dynamometer can help you fine-tune the ignition timing for maximum performance and efficiency.

Cooling system failures can also impact your engine's performance during tuning. Symptoms might include overheating, loss of power, or coolant leaks. Start your troubleshooting by inspecting the radiator, hoses, and water pump for any visible signs of wear or damage. Ensure that the coolant levels are adequate and that the thermostat is functioning properly. If you've made modifications to increase engine performance, consider upgrading to a more efficient cooling system to prevent overheating, especially during demanding driving conditions or track days.

Finally, suspension issues can affect your Subaru's handling and overall performance. If you notice excessive body roll, uneven tire wear, or a rough ride, it may indicate problems with your suspension components. Check the struts, shocks, and bushings for signs of wear or damage, and ensure that your alignment is set correctly. Upgrading to performance suspension parts can also enhance your vehicle's handling characteristics, but it's essential to ensure that these modifications are compatible with the rest of your tuning setup.

In conclusion, troubleshooting common issues in Subaru performance tuning requires a systematic approach to diagnosing and resolving problems. By becoming familiar with the typical symptoms associated with air-fuel mixture, ignition timing, cooling systems, and suspension, you can effectively address these challenges. Regular maintenance and a proactive mindset will not only help to keep your Subaru running smoothly but will also enable you to enjoy the full benefits of your performance modifications. With persistence and the right knowledge, you can overcome these hurdles and truly master the art of Subaru performance tuning.

Chapter 8: Safety Considerations

Essential Safety Equipment

When delving into the world of performance tuning and engine modifications, it's easy to become enamored with the thrill of increased horsepower, improved handling, and the aesthetics of a finely-tuned Subaru. However, while enhancing your vehicle's performance, it is crucial to prioritize safety. Essential safety equipment not only protects you and your passengers but also safeguards your investment in your car. Understanding what gear is necessary can mean the difference between a thrilling experience and a disastrous one.

First and foremost, a quality helmet is non-negotiable for any automotive enthusiast venturing onto the track or engaging in spirited driving. A helmet designed specifically for motorsport use provides superior protection against impacts and can significantly reduce the risk of head injuries. Look for models that meet or exceed the Snell or DOT safety standards, which indicate rigorous testing for performance and durability. Additionally, consider features like ventilation and weight, as a comfortable helmet will encourage you to keep it on during long sessions.

Equally important is the use of a proper racing harness. In high-performance situations, the standard seatbelt may not offer the level of restraint needed to keep the driver firmly in place during hard cornering or sudden stops. A racing harness, typically a five-point or six-point system, distributes forces more evenly across the body and minimizes movement, which is vital in maintaining control of the vehicle. When selecting a harness, ensure it's compatible with your seats and meets safety standards set by organizations such as the SFI or FIA.

Fire safety equipment is another critical component of any performance tuning setup. A fire extinguisher rated for automotive use should be easily accessible within the car. Ideally, it should be securely mounted in a location where it can be quickly retrieved in case of an emergency. Additionally, consider investing in a fire suppression system, especially for those who frequently participate in motorsport events. These systems can rapidly extinguish flames and significantly increase your chances of escaping a vehicle fire unscathed.

Lastly, don't overlook the importance of proper attire. While it may seem trivial, wearing flame-resistant clothing and gear can protect you in the event of a fire. Racing suits, gloves, and footwear designed for motorsport use are made from materials that resist ignition and help prevent burns. Additionally, ensure your footwear provides adequate grip and support, as losing control due to improper shoes can lead to accidents. By equipping yourself with the right safety gear, you not only enhance your driving experience but also reinforce the commitment to safety that every automotive enthusiast should uphold.

Engine and Component Reliability

Engine and component reliability is a cornerstone of performance tuning, particularly when it comes to Subaru vehicles. As enthusiasts delve into modifications, it becomes paramount to understand how these changes can affect the longevity and durability of the engine and its components. Subaru engines, known for their robust design and performance capabilities, can withstand a fair amount of tuning, but each modification introduces variables that can compromise reliability if not approached with care and knowledge.

When tuning a Subaru, one of the first considerations is the engine's ability to handle increased power outputs. The factory specifications are designed to provide a balance between performance and reliability, but pushing beyond these limits requires a solid understanding of the engine's components. Key areas such as the pistons, connecting rods, and crankshaft must be evaluated for their strength and performance under enhanced loads. Upgrading these components can help mitigate the risks associated with high-performance tuning, ensuring that the engine remains reliable even when subjected to greater stresses.

In addition to internal components, the reliability of ancillary systems like fuel delivery, cooling, and exhaust also plays a critical role. Modifications to the engine often necessitate upgrades to the fuel injectors, fuel pumps, and intercoolers to maintain optimal performance. For instance, an upgraded fuel pump may be required to support a higher horsepower setup, but if not paired with a suitable fuel management system, it can lead to lean conditions that jeopardize engine health. Similarly, ensuring that the cooling system is capable of dissipating increased heat generation is vital for preventing overheating and potential engine failure.

Another aspect to consider is the importance of tuning software and engine management systems. Modern Subarus are equipped with sophisticated engine control units (ECUs) that require recalibration when significant modifications are made. A poorly executed tune can lead to issues such as knock, misfires, or excessive exhaust temperatures, all of which can harm engine longevity. Therefore, working with experienced tuners who understand the intricacies of Subaru's engine management is essential for achieving a reliable setup that maximizes performance without compromising durability.

Ultimately, the key to achieving enhanced performance without sacrificing reliability lies in a holistic approach to tuning. This involves not only upgrading critical components but also ensuring that all systems work in harmony. Regular maintenance, thorough testing, and a commitment to understanding the limits of each modification will lead to a performance vehicle that not only delivers exhilarating power but also stands the test of time on the road or track. For Subaru enthusiasts, this balance is the essence of mastering performance tuning, ensuring that their boosted dreams become a thrilling reality without the nightmare of engine failure.

Legal Considerations in Tuning

Tuning a Subaru for enhanced performance can be an exhilarating journey, but it is essential to navigate the legal landscape that accompanies modifications. This subchapter aims to provide automotive modders and tuners with a comprehensive understanding of the legal considerations involved in tuning. From emissions regulations to warranty implications, being informed about these aspects can help ensure that your boosted dreams remain within the bounds of the law.

One of the most critical legal considerations in performance tuning is compliance with emissions regulations. In many regions, vehicles must adhere to specific emissions standards set forth by governmental agencies. Modifications that affect the exhaust system or engine performance can potentially lead to increased emissions, which may render the vehicle illegal for street use. It is crucial to research local laws and regulations, as these can vary significantly from one jurisdiction to another. For instance, modifications that might be acceptable in one state could lead to fines or even the inability to register the vehicle in another.

Another aspect to consider is the impact of tuning on vehicle warranties. Many manufacturers, including Subaru, have stringent policies regarding modifications. Altering the engine or related systems may void existing warranties, leaving the owner responsible for any subsequent repairs. This risk is particularly pertinent for those who drive newer models or still have active warranties. It's advisable to consult the warranty documentation and, if necessary, speak with the dealership to understand the implications of specific modifications before proceeding with any tuning work.

Insurance considerations also play a vital role in the legal landscape of performance tuning. Modifying a vehicle can affect its insurance coverage and rates. Insurers may classify a modified vehicle as a higher risk, resulting in increased premiums or even denial of coverage altogether. Additionally, it is crucial to inform your insurance provider of any modifications to ensure that you remain protected in the event of an accident or theft. Failing to disclose modifications could lead to complications when filing a claim, leaving you vulnerable to financial repercussions.

Finally, understanding the legal ramifications of street racing and other performance-related activities is essential for any automotive enthusiast. Engaging in illegal street racing can lead to severe penalties, including fines, impounding of the vehicle, and even criminal charges. To stay within the law, consider participating in organized motorsport events that provide a safe and legal environment for showcasing your tuning skills. By respecting the rules of the road and exploring legal racing venues, you can enjoy the thrill of performance tuning while minimizing legal risks.

In summary, while the pursuit of performance tuning can be an exciting venture, it is crucial to remain mindful of the legal considerations that accompany it. By understanding emissions regulations, warranty implications, insurance requirements, and the laws surrounding street racing, automotive modders and tuners can navigate the complexities of tuning with confidence. Staying informed and compliant will not only protect your investment but also ensure that your journey toward mastering Subaru performance tuning remains a fulfilling and enjoyable experience.

Chapter 9: Community and Resources

Engaging with the Subaru Community

Engaging with the Subaru community is an essential aspect of enhancing your experience as an automotive modder and tuner. The Subaru enthusiast culture is vibrant and multifaceted, with a rich history that dates back to the brand's rally roots. Being part of this community not only provides access to a wealth of knowledge and resources but also fosters connections with like-minded individuals who share your passion for performance tuning and engine modifications. This subchapter explores various avenues for engaging with fellow enthusiasts, from online forums to local meetups, ensuring you tap into the collective wisdom of the Subaru family.

Online forums and social media platforms serve as the backbone of the Subaru community, offering a space for enthusiasts to exchange ideas, troubleshoot issues, and showcase their builds. Websites like NASIOC (North American Subaru Impreza Owners Club) and Reddit's r/Subaru are treasure troves of information, featuring discussions ranging from basic maintenance tips to advanced tuning techniques. By actively participating in these platforms—asking questions, sharing your experiences, and offering advice—you not only enhance your own knowledge but also contribute to the community's growth. Engaging in these discussions can lead to new friendships and collaborations, as well as potential opportunities for group buys or joint projects.

Local car meets and Subaru clubs provide another layer of engagement that can significantly enrich your automotive journey. These events allow you to see a variety of Subaru models up close, giving you inspiration for your own build. Meeting fellow enthusiasts face-to-face fosters camaraderie and opens doors to sharing experiences, tips, and techniques that may not be extensively covered online. Many clubs also organize events like track days, rallycross, or dyno days, which can offer invaluable hands-on experience and a chance to test your modifications in a controlled environment.

Attending Subaru-specific events, such as the annual Subaru Summer Solstice or the Subiefest, can deepen your connection to the community. These gatherings attract enthusiasts from all over, showcasing not only modified vehicles but also the latest aftermarket products and tuning technologies. They often feature workshops and seminars led by experts in the field, providing insights into performance tuning and engine modifications. Participating in such events allows you to stay abreast of trends and innovations while also networking with industry professionals and other enthusiasts who share your passion.

Finally, engaging with the Subaru community is about giving back. Sharing your own experiences, whether they are successes or failures, helps to create a supportive environment where everyone can learn. Documenting your tuning journey through blogs or social media can inspire others and provide valuable insights that contribute to the collective knowledge base. By mentoring newcomers or collaborating on projects, you not only enhance your own skills but also strengthen the community as a whole. Ultimately, engaging with the Subaru community enriches your experience as a modder and tuner, fostering a sense of belonging and shared passion that transcends individual projects.

Online Resources and Forums

In the ever-evolving world of automotive performance tuning, particularly for Subaru enthusiasts, online resources and forums have become invaluable tools. These platforms provide a wealth of information, from technical specifications to community-driven advice, enabling modders and tuners to enhance their vehicles effectively. Whether you're a novice seeking guidance or an experienced tuner looking for the latest trends, the internet is teeming with resources that can help elevate your Subaru to new heights.

One of the most significant advantages of online forums is the sense of community they foster. Subaru enthusiasts from all walks of life come together to share their experiences, successes, and failures. This collective knowledge is often more practical than any manual or guide, as it draws upon real-world applications and solutions to common problems. Users can post questions about specific modifications, troubleshooting issues, or even seek recommendations for aftermarket parts. The camaraderie found in these forums can be a great motivator, encouraging members to push the limits of their vehicles while learning from one another.

Additionally, many online resources provide detailed guides and articles focusing on performance tuning and engine modifications. Websites dedicated to Subaru tuning often feature step-by-step instructions, covering everything from basic upgrades like cold air intakes to more complex modifications such as turbocharger installations. These guides frequently include diagrams, photos, and even video tutorials that demystify the tuning process. By leveraging these resources, tuners can develop a deeper understanding of how various components interact and how to optimize their vehicle's performance effectively.

Social media platforms also play a pivotal role in connecting Subaru enthusiasts. Groups and pages dedicated to performance tuning provide a space for real-time discussions, showcasing builds, and sharing tips. Instagram and Facebook, in particular, serve as visual platforms where tuners can display their projects, engage with peers, and gain inspiration from others' work. This dynamic interaction not only helps modders stay updated on the latest trends and modifications but also allows them to receive immediate feedback on their ideas and projects.

Lastly, it's essential for automotive modders to remain vigilant about the credibility of the information they encounter online. With the vast amount of content available, discerning between reliable sources and those that may lead to costly mistakes is crucial. Engaging with well-established forums and resources, verifying information through multiple channels, and leaning on community feedback can significantly reduce the risk of misinformation. By harnessing the power of online resources and forums wisely, Subaru enthusiasts can not only improve their vehicles' performance but also become part of a vibrant, knowledgeable community dedicated to the art of tuning.

Attending Events and Meets

Attending events and meets is a pivotal aspect for automotive enthusiasts, particularly for those involved in performance tuning and engine modifications. These gatherings provide a unique opportunity to immerse oneself in a community of like-minded individuals who share a passion for Subaru vehicles. From local car meets to larger automotive events, these occasions serve as platforms for showcasing customized vehicles, sharing knowledge, and fostering friendships that can last a lifetime. Understanding the significance of attending these events can greatly enhance your tuning journey and help you stay up-to-date with the latest trends and techniques in the industry.

One of the primary benefits of attending Subaru events is the chance to network with other enthusiasts and experts in the field. Engaging in conversations with fellow modders can lead to valuable insights about tuning techniques, parts compatibility, and performance upgrades. Whether you're looking to optimize your vehicle's turbo setup or seeking advice on suspension modifications, the collective experience of the community can prove invaluable. Many seasoned tuners are eager to share their knowledge, and building these relationships can open doors to collaboration on future projects or even sponsorship opportunities.

In addition to networking, attending meets allows for direct observation of various tuning styles and modifications. Each vehicle is a testament to the owner's creativity and technical abilities, providing inspiration for your own projects. Observing how others have approached performance enhancements can spark new ideas and lead to innovative solutions for challenges you may face.

Furthermore, attending events often presents the chance to see cutting-edge products and technologies up close, allowing you to make informed decisions when selecting parts for your Subaru.

Events and meets also serve as a platform for learning through workshops and demonstrations. Many gatherings feature presentations from industry experts who can provide in-depth knowledge about performance tuning, engine management systems, and the latest in aftermarket products. Participating in these sessions can deepen your understanding of the principles behind tuning and help you refine your skills. The hands-on experience gained from workshops can be particularly beneficial, as it allows you to apply theoretical knowledge in a practical setting, enhancing your overall proficiency as a tuner.

Finally, attending events fosters a sense of belonging within the Subaru community. The shared enthusiasm for performance tuning and engine modifications creates an environment of camaraderie that transcends geographical boundaries. Whether you're a novice modder or a seasoned professional, the support and encouragement found at these gatherings can be incredibly motivating. Joining in on group activities, such as track days or show-and-shine competitions, can not only elevate your skills but also solidify friendships that extend beyond the automotive realm. In essence, attending events and meets is not merely an avenue for showcasing your Subaru; it is an integral part of the journey towards mastering the art of performance tuning.

Chapter 10: The Future of Subaru Performance Tuning

Emerging Technologies and Trends

Emerging technologies and trends in the automotive world are reshaping the landscape of performance tuning and engine modifications, especially for Subaru enthusiasts. As the demand for higher performance and more efficient vehicles grows, innovative solutions are continuously being developed. This subchapter explores some of the most exciting advancements in technology that are set to transform how modders and tuners approach Subaru performance enhancement.

One of the most significant trends is the rise of digital tuning tools and software. Advanced ECU tuning software has become increasingly user-friendly, allowing tuners to access and modify engine parameters with unprecedented ease. Platforms like Cobb Accessport and Ecutek have become staples in the Subaru community, providing detailed data logging and real-time adjustments. These tools not only optimize performance but also enhance the ability to diagnose issues and fine-tune setups on the fly. As these technologies evolve, they promise to offer even deeper insights into engine management and performance optimization.

Another noteworthy development is the integration of artificial intelligence (AI) in performance tuning. AI algorithms can analyze vast amounts of data from a vehicle's performance metrics, enabling tuners to make more informed decisions. For instance, AI-driven systems can predict how changes in one parameter will affect overall performance, allowing for smarter modifications. This technology can lead to more efficient tuning processes, reducing the trial-and-error approach that often characterizes performance upgrades. As tuners begin to leverage AI, we can expect a shift towards data-driven decision-making in Subaru modifications.

The use of lightweight materials is also gaining traction within the Subaru modding community. Manufacturers are increasingly incorporating materials such as carbon fiber, aluminum, and advanced composites into their vehicles, which can significantly reduce weight and improve handling. For tuners, this trend opens up new possibilities for enhancing performance without solely relying on engine modifications. Upgrading components like hoods, bumpers, and even wheels with lighter alternatives can lead to improved acceleration, braking, and overall dynamics. As these materials become more accessible, modders will likely adopt them to create even more capable Subaru builds.

Additionally, the electric and hybrid technology wave is starting to influence the performance tuning sphere. While traditional Subaru models are known for their internal combustion engines, the integration of electric assist systems presents a new frontier for performance enhancement. Hybrid setups can provide immediate torque and improve overall efficiency, offering tuners innovative ways to boost performance without compromising on fuel economy. As manufacturers explore electrification, Subaru enthusiasts may find themselves adapting their tuning strategies to incorporate these emerging technologies, leading to a new era of hybrid performance tuning.

In conclusion, the automotive world is in a state of rapid evolution, with emerging technologies and trends offering exciting opportunities for Subaru modders and tuners. From advanced digital tuning tools and AI-driven insights to lightweight materials and hybrid technology, the landscape of performance tuning is transforming. As these advancements continue to develop, Subaru enthusiasts will need to stay informed and adaptable, ensuring they remain at the forefront of performance enhancement in the ever-changing automotive environment. Embracing these technologies will not only enhance individual builds but also contribute to the broader community of performance tuning, paving the way for future innovations.

Electric and Hybrid Performance Tuning

Electric and hybrid performance tuning represents a significant evolution in the automotive landscape, particularly for Subaru enthusiasts who have traditionally focused on internal combustion engines. As manufacturers increasingly develop hybrid and electric powertrains, understanding how to optimize these systems is essential for tuners looking to push the boundaries of performance. This subchapter delves into the unique challenges and opportunities presented by electric and hybrid vehicles, offering valuable insights for those ready to embrace this new frontier.

Electric vehicles (EVs) are powered by electric motors, which operate differently from conventional engines. The instantaneous torque delivery of electric motors offers instant acceleration, making them inherently responsive. Tuners can enhance this performance by modifying the vehicle's software to adjust parameters such as throttle response and power delivery curves. Many modern EVs, including hybrids, come equipped with sophisticated onboard computers that control various aspects of performance, from energy management to regenerative braking. By accessing these systems through specialized tuning software, enthusiasts can fine-tune the driving experience to suit their preferences, whether that involves maximizing power output or enhancing efficiency.

Hybrid vehicles, which combine internal combustion engines with electric motors, present their own set of tuning opportunities. The integration of both power sources allows for unique performance strategies, such as optimizing the balance between electric and gasoline power. Tuners can modify the engine control unit (ECU) to improve the hybrid system's efficiency, thereby enhancing performance without sacrificing fuel economy. Additionally, adjusting the parameters governing the electric motor's engagement can result in notable improvements in acceleration and overall drivability. This dual approach allows Subaru modders to leverage the strengths of both technologies, creating a seamless and exhilarating driving experience.

Moreover, performance tuning for electric and hybrid vehicles often includes enhancing the vehicle's thermal management system. Electric motors and batteries generate heat, and efficient cooling is crucial for maintaining optimal performance. Upgrading cooling systems or integrating aftermarket solutions can prevent overheating, thus allowing for sustained performance during aggressive driving. Additionally, tuners can explore the use of lightweight materials and aerodynamic modifications to improve overall efficiency, further enhancing the performance capabilities of these vehicles.

As the automotive industry continues to evolve towards electrification, Subaru enthusiasts must adapt their tuning skills to incorporate these new technologies. The performance tuning of electric and hybrid vehicles is not just about increasing horsepower; it requires a comprehensive understanding of how these systems work together. By embracing the challenges and opportunities presented by electric and hybrid performance tuning, Subaru modders can continue to push the limits of what is possible, creating innovative and high-performing machines that reflect their passion for automotive excellence.

Final Thoughts on Your Subaru Journey

As you reach the conclusion of your journey into the world of Subaru performance tuning, it's essential to reflect on the experiences and knowledge you've gained along the way. The Subaru community is more than just a collection of enthusiasts; it's a family that shares a passion for enhancing performance, embracing innovation, and enjoying the thrill of the ride. Whether you've modified your Subaru for daily driving, competitive racing, or simply for the love of the craft, every step of this journey has contributed to your growth as a modder and tuner.

Understanding the intricacies of Subaru's engineering is critical when embarking on performance modifications. From the iconic boxer engine layout to the advanced all-wheel-drive systems, Subaru vehicles are designed with performance in mind. This engineering foundation provides ample opportunities for tuning, allowing you to explore various modifications that enhance power, handling, and overall driving dynamics. As you progress, remember that each modification should align with your specific goals, whether they involve increasing horsepower, improving throttle response, or refining suspension setup for better cornering capabilities.

As you experiment with tuning, it's vital to emphasize the importance of a balanced approach. Performance tuning is not just about raw numbers; it's about creating a vehicle that performs harmoniously across all parameters. This means considering factors such as weight distribution, brake upgrades, and tire selection. An effective tuning strategy will integrate all these elements to create a cohesive driving experience. By taking the time to evaluate each aspect of your Subaru, you can achieve not just a powerful machine, but also one that is enjoyable and reliable on the road or track.

Safety should always be a top priority in your tuning journey. As you push the limits of your Subaru's performance, ensure that you also invest in safety upgrades. This includes quality brakes, upgraded suspension components, and the installation of safety features such as harnesses and roll cages if you plan to take your vehicle to the track. Being mindful of safety will not only protect you but also enhance your confidence in pushing your vehicle to its limits.

In closing, your Subaru journey is an ongoing adventure filled with learning, experimentation, and camaraderie. The knowledge you've gained through this book and your experiences in the community will serve as a solid foundation for continued exploration in performance tuning. As you modify and enhance your Subaru, remember to share your insights and celebrate the achievements with fellow enthusiasts. Each journey is unique, but together, we can elevate the Subaru experience and continue to push the boundaries of what's possible in automotive performance.

www.ingramcontent.com/pod-product-compliance
Lightning Source LLC
Chambersburg PA
CBHW070417230526
45471CB00006B/2845